Welcome to Spooky Specimens! I'm glad you're here :)

I'm going to be introducing you to 31 creatures – some alive and some extinct – that I believe deserve the title of "spooky" (well, technically, there's one "creature" in here that is more of a place than a thing).

To me, "spooky" can mean a few different things. Spooky can be "kinda cute, but kinda creepy", and it can also be "godforsaken", or "abysmal", or maybe "foul". You can decide which ones are which.

Along with the pictures of each specimen for you to color, you'll have a short description of each one to read (some of them ended up being longer than I expected...I had a lot to say). You'll probably end up horrified or educated. Or maybe both.

I can't wait to see how you bring these to life :)

Brazilian Yellow Scorpion

Scorpions are already the stuff of nightmares, but the Brazilian yellow scorpion is a hell-sent creature. At only 2-3 inches long (5-7 cm), they're the most dangerous scorpion – in recent years, arguably the most dangerous animal – in South America. Their venom is extremely toxic, leading to thousands of hospitalizations and on rare occasions, some fatalities.

What makes them so dangerous is not necessarily the potency of their venom, but the rate at which they're infiltrating cities. Turns out they're perfectly adapted for city life. Their favorite food is cockroaches (cities have an endless supply of them), but an abundance of their prey is not even necessary for their survival. They can go 400 days without food! Females can also reproduce without a mate through a process called parthenogenesis, where they fertilize their own eggs. She can produce 30 offspring this way, multiple times a year. It's like they have cheat codes for survival.

Now, they're everywhere: lurking in the shadows and crawling up kitchen sinks. Someone in Brazil even came across an abandoned house with thousands of them sprawled across the floors and the walls.

The number of envenomations in recent years reveals just how fast they're invading cities. In 2000, Brazil saw 12,000 cases across the country. In 2018, that number increased to 156,000. Mild side effects include nausea, vomiting, and shortness of breath; severe side effects may be myocardial damage, cardiac arrhythmias, excess fluid in the lungs, and death.

Humboldt Squid

The Humboldt squid, or "the red devil", is a large and extremely predatory squid found in the deeper waters of the eastern Pacific Ocean (660–2,300 ft / 200–700 m deep). They can get up to 8 ft (2.4 m) long and 100 lbs (45 kg). Like many cephalopods, they're covered in chromatophores (cells that produce color), which allow them to change color to communicate. When aggravated, they become bright red (hence the nickname "red devil"), and have been known to act aggressively towards scuba divers on rare occasions.

You might be thinking, "Lindsay, there are other giant squid species that get much larger than these. Why did they get put into this book? What makes them spookier than the others?"

Alone, their spooky factor might be a bit low. But the Humboldt squid is known to live – and hunt – in groups.

Up to 10? No.

Up to 100? No.

In groups of over one thousand.
That makes me sick to my stomach.

When hunting in these groups, they're able to go after larger prey and are known to drag their victims into the depths so that they go unconscious.

Goliath Bird Eater

Let's cut right to the chase. This is the largest spider to ever exist (that we know of) and we just so happen to live at the same time as them. There's an extinct arthropod that used to hold this title (Megarachne), but it has now been reclassified as a sea scorpion. So our guy, the bird eater, currently reigns supreme. They reside in South America.

Including their legs, they can get to about a foot long (30 cm). Maybe you're thinking "that's not as bad as I thought it would be," or maybe you're more of a "that's absolutely vile, I don't want to share a planet with a creature like that" type. Either way is valid.

Despite their name, they only eat birds a little bit. They mostly eat insects. But they have also been known to eat frogs and rodents. They use their inch-long fangs to inject neurotoxins into their prey. They then drag their victim back to its burrow and suck out their insides.

Covering the entire surface area of their legs are modified leg hairs to sense vibrations in their environment. These leg hairs also happen to have little stinging barbs at the end of them. They use these as a defense mechanism in a peculiar way. They rub their legs together (a motion that creates a hissing noise) and the barbs are shot into the air, hopefully landing somewhere on the threat to send it running off.

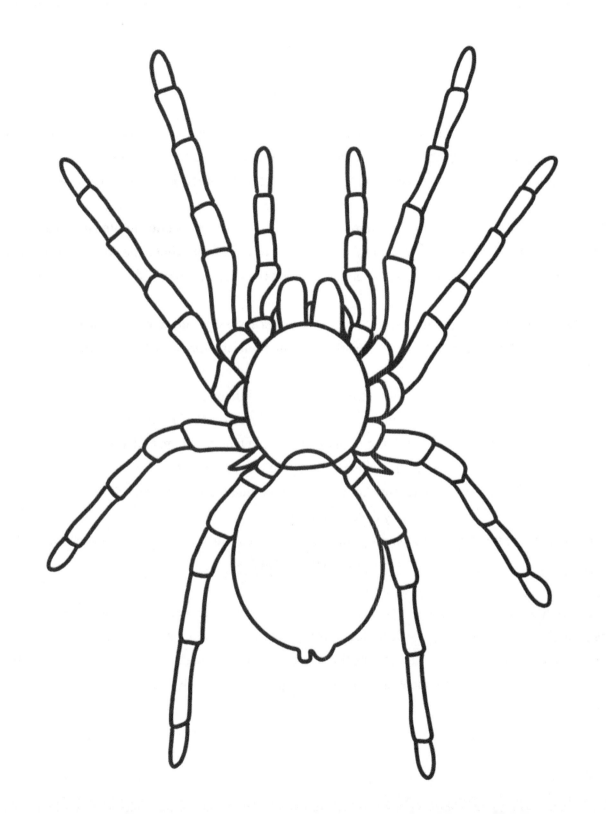

Hammerhead Worm

If you've ever taken a biology class, you've probably heard of a planarian: a type of flatworm with amazing regenerative abilities. They're covered in stem cells, which allow them to regrow any part of their body after being sliced and diced – an exciting specimen for research. Most planarians are 3-15 mm long.

Now imagine that a mad scientist took one of these small, cool little guys and mutated it over and over again until it became a massive, monstrous version of its original form. That's the hammerhead worm. Except the mad scientist is millions of years of evolution.

Recently, hammerhead worms have been a huge topic of discussion, mainly due to four key characteristics: they're invasive, toxic, cannibalistic, and potentially immortal. I'll break it down.

Invasive: They're originally from Southeast Asia but due to human travel/transportation, now have a worldwide distribution. They have the power to destroy ecosystems by eradicating earthworms (their favorite food). They trap earthworms in this slime that liquefies them and they drink it up.
Toxic: The slime is toxic. Don't try to eat them.
Cannibalistic: Hammerhead worms are known to eat other hammerhead worms. When food is scarce, they'll also eat parts of themselves.
Immortal: As I mentioned at the beginning, they are planarians, and have the same regenerative abilities as their smaller cousins.

Since they're an invasive species, ecologists actually recommend killing these worms if you see them. Obviously, you can't chop them into little pieces. You'll just get more hammerhead worms. The only way to ensure they'll actually die, without harming the surrounding environment (with, say, fire), is by trapping them in a jar with salt and/or vinegar until they fully dissolve.

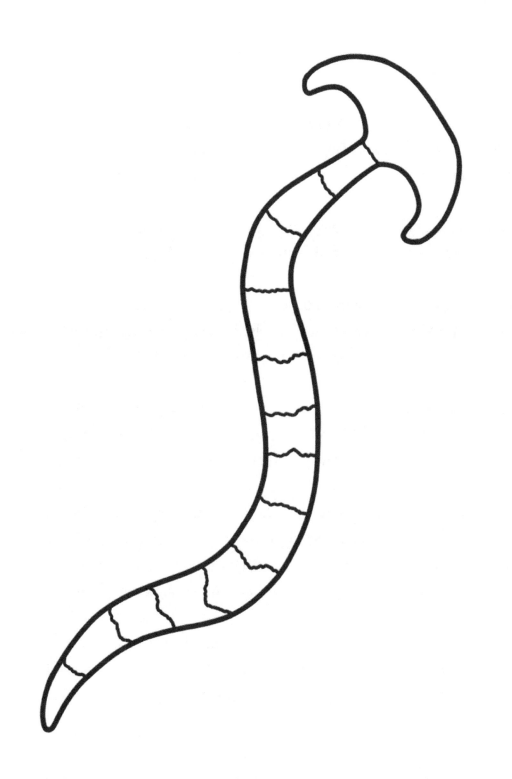

Titanoboa

66 million years ago, a 7-mile (11 km) wide asteroid hurtled into the southern part of the Gulf of Mexico, leading to the extinction of all of the non-avian dinosaurs and 75% of all life on Earth. The species that survived were small, and they remained this way for millions of years.

Until Titanoboa stepped on the scene. The largest snake to ever exist (that we know of) and the largest predator of its time (that we know of) 60 million years ago.

Fossil evidence suggests they could get to 50 ft (15 m) long. If you're having trouble picturing that, it's about the length of a semi-truck trailer. It's also the length of each letter of the Hollywood sign! (That one's more of a fun fact.)

What really freaks me out, though, is how wide they could get. Three. Feet. Wide. (1 m). To picture that, imagine a massive snake that could barely squeeze through your bedroom door, and would come up to your waist while slithering completely flat on the ground.

So how did they get so big? Well, snakes are ectotherms, meaning they get their body heat from the external environment (compared to us who are endotherms, and produce our own body heat). Their environment, of what is now Colombia, was much warmer than it is today. And with the way their metabolism works, warmer temperatures allow them to get larger.

Like modern-day anacondas, Titanoboa was likely a constrictor, able to take down large prey such as crocodiles. You should be grateful they're not around today, as you would probably make for a perfectly-sized snack.

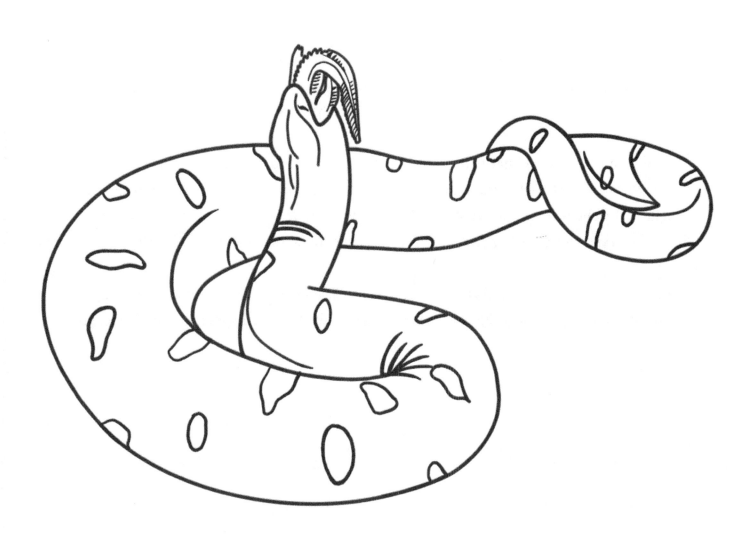

Bigfin Squid

Ugh. I wish I had more to say about these guys. They're one of the many mysterious creatures found in the deep sea – rarely seen – so we don't know much about them. So far, there's been fewer than 20 verified sightings.

What we do know is that they're typically 13 - 26 ft (4 - 8 m) long, and most of that length comes from their 8 spaghetti-like arms and 2 tentacles. They sometimes hold them at a perpendicular angle to their bodies, creating what looks like elbows. They've been spotted as deep as 16,000 ft (4800 m), so luckily you've never come across them swimming at the beach.

Their appendages have microscopic suckers on them. Scientists aren't exactly sure how they use them, but they suggest they might hang them, dragging them along the seafloor, waiting for prey to bump into them.

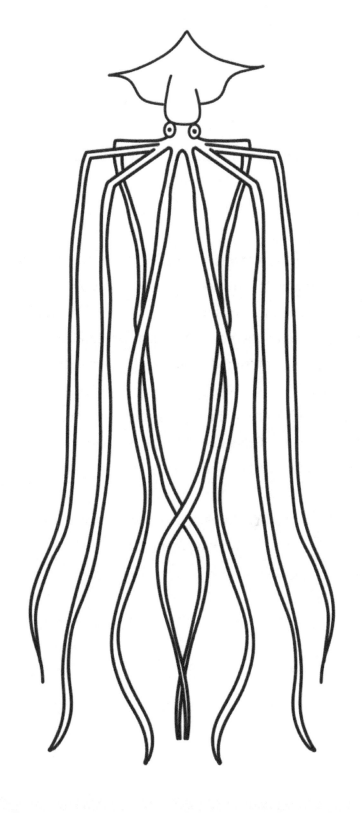

Brine Pools

These are arguably the most mysterious habitats on the planet. They're underwater lakes found in pockets of the deep sea. If you were a fish, this might seem like a cool, maybe inviting place to visit (like Goo Lagoon??). But think again. Brine pools are death traps.

The pools are extremely salty concentrations of water – so thick that the salt can't dissolve with the surrounding (regular) salt water. It just sticks together in this oversaturated blob. Since these concentrations are heavier than the surrounding water, they sink, creating what looks like the surface of a pool or a lake. Underneath that surface is a thick liquid seeping with toxic chemicals and no oxygen, sending animals that dare enter it into toxic shock. And then they die.

Some animals have learned their limits in their pools, and play a dangerous game to scavenge the carcasses of the unfortunate victims. Others have adapted to live right on the edge, testing our own understanding of the limits of life on Earth.

There are some very cool videos on YouTube of ROVs entering brine pools. I highly recommend checking them out.

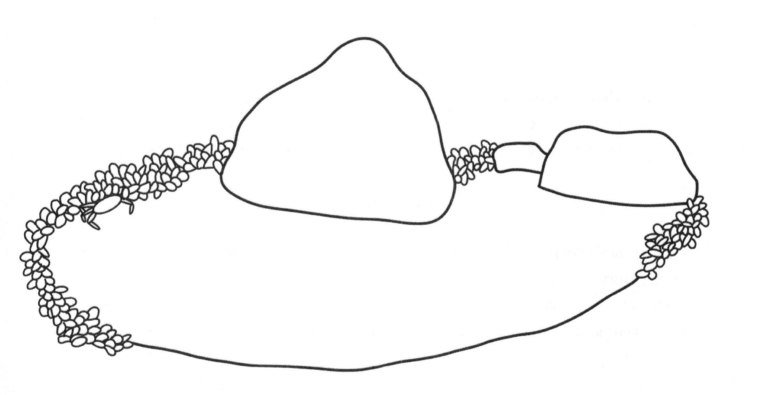

Anglerfish Mating

This one is for the lovebugs <3

You might be having a difficult time figuring out what's going on in this picture. Those are not two legs with eyes on them. Those are male anglerfish.

Anglerfish have quite a unique way of mating. Male anglerfish are much smaller than females – they don't even look like the same species. They have large eyes and nostrils to detect chemicals that female anglerfish emit as an attractant, to fulfill their only purpose in life.

Giving females sperm.

When a male anglerfish finally finds his female, he bites down on her and doesn't let go. Ever. Over time, his tissues and circulatory system fuse with hers. He slowly loses his eyes, teeth, most of his internal organs, and all sense of individuality and purpose. He becomes a "sexual parasite", surviving on nutrients from the female's blood, and providing sperm for her whenever she is ready.

Spider Dance of Death

For male spiders, mating is like a game of Russian roulette. The wrong move could cost you your life.
Different species have different mating rituals. I figured the peacock spider would be the most fun to color in.

For male peacock spiders, they have to dance like their life depends on it (and it literally does.). They move their entire body around for the chance of mating, throwing their appendages into the air. A hungry female might pounce in to try to kill the male. If he's swift enough, he'll work it into his dance moves to woo her.

Males of other species have different tactics. A black widow (along with many other species) will vibrate his abdomen on a female's web, sending vibrations through the silk strands to indicate his presence. The wrong vibrations might indicate he's prey instead of a potential mate, and he will be eaten. Male nursery web spiders bring gifts to the female wrapped in silk. What's inside? The trophy carcasses of his prey! If she doesn't like it or she's having a bad day, she'll eat him.

Vampire Squid

Neither squid nor octopus, the vampire squid belongs to its own cephalopod order: Vampyromorphida. Their scientific name is Vampyroteuthis infernalis, which translates to "vampire squid from hell." With a name like that , you'd expect this creature to terrorize the deep sea, slandering anything that crosses its path. That's the furthest from the truth. The vampire squid is a passive, soft-bodied creature.

Their cape-like arms and red coloration are the characteristics that gave them their name (I added vampire teeth for fun. Those are unfortunately not real). They're about a foot long (30 cm) and are found at depths of 2000-3000 ft (600-900 m). Their "ears" are actually fins! They help them move through the water.

Vampire squids eat drifting particles called "marine snow." When threatened, they expel a colorless substance with bioluminescent particles (in contrast to dark ink) to confuse the threat.

They are definitely "spooky cute." :)

Goblin Shark

In contrast to the vampire squid, we can understand exactly why the goblin shark got its name. It's freaky looking. No offense.

You've probably seen videos online of their jaws fully ejecting from their braincase (isn't that a great word?) in a process called "slingshot feeding." Their upper and lower jaws lunge forward, away from the skull, engulfing their prey.

This process might seem alien, but it's actually not uncommon for fish. One might argue the goblin shark has the most "extreme" case of this process, but I would actually disagree. I think the title goes to the slingjaw wrasse (Epibulus insidiator). Look it up online. Right now.

The goblin shark is found at depths of up to 4265 ft (1,300 m) below the surface, and they can get up to 12 ft long. They eat a varied diet of squid, fish, and crustaceans after locating them with sensory organs all over their snout.

Executioner Wasp

The executioner wasp gained tons of attention after Coyote Peterson informally crowned this species as the most painful sting out there. So I decided to use the wasp as the ambassador species for this page: level 4 of the sting pain index.

The sting pain index was created by entomologist Justin Schmidt to rank the relative pain of different insect stings. You might think this sounds insane. I'd consider it brave science.
Someone had to do it, right?

He subjected himself to over 80 different stings and placed them on a rating scale of 1 - 4 (4 being beyond excruciating). Here's how he described the three species he placed at a 4 on his index:

Tarantula Hawk Wasp: "Blinding, fierce, shockingly electric.
A running hair dryer has just been dropped in your bubble bath."

Warrior wasp: "Torture. You are chained in the flow of an active volcano. Why did I start this list?"

Bullet Ant: "Pure, intense, brilliant pain. Like walking over flaming charcoal
with a 3-inch nail embedded in your heel."

Recently, a man named Coyote Peterson decided to follow in Schmidt's footsteps, but he also ended up subjecting himself to the sting of a newly discovered species Schmidt never experienced:
the executioner wasp. They're found in Central and South America, and apparently have a much more painful sting than the bullet ant (whom Schmidt and Coyote had previously crowned supreme). The sting burned a hole in his arm.

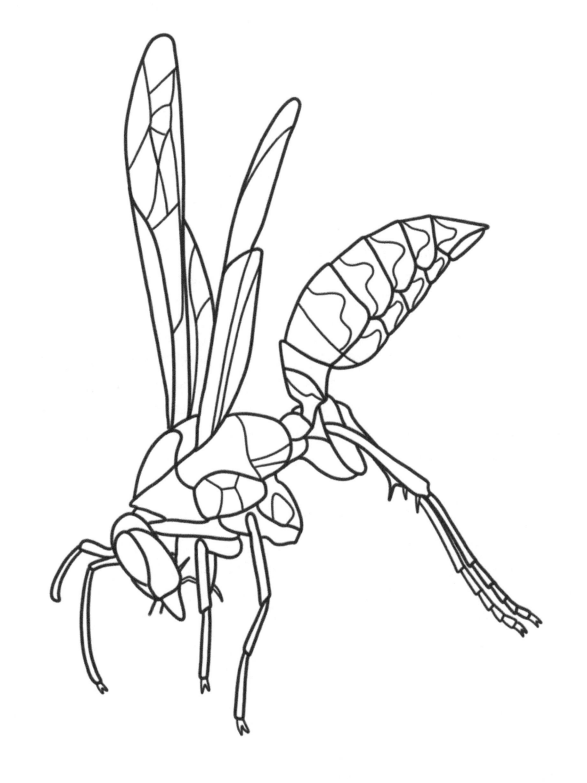

Tully Monster

This is the weirdest extinct animal ever discovered. Scientists have been trying to figure out what it even is, and what it's related to, for decades. It's called Tullimonstrum, or the tully monster, and it was alive 300 million years ago.

The tully monster has some very strange and unique characteristics. They have a trunk-like snout with a claw at the end. That's their mouth. It has two rows of teeth inside of it. They also have eyes on stalks that extend sideways (like wtf). Both traits haven't been seen on any other animal.

For a long time, it was thought the tully monster was an invertebrate (an animal without a backbone). Then, in 2016, scientists seemed to solve the mystery by identifying vertebrate characteristics using over 1,200 fossils. They also seemed to determine that their closest living relatives are lampreys (a type of jawless fish). However, many scientists disagree with this classification and still think there's tons of evidence pointing to them being invertebrates.

So for now, the tully monster remains a mystery.

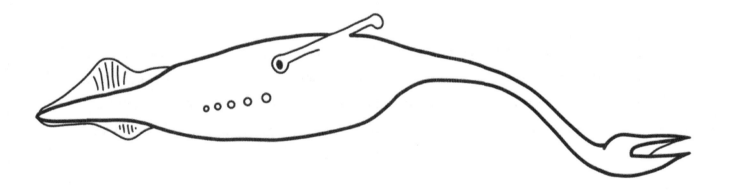

Pink Dragon Millipede

The pink dragon millipede is exactly how it sounds: hot pink. Very cute. They were only recently discovered in 2007 in Thailand and are about 3 cm long.

Like many animals, their bright coloration acts as a warning signal (the scientific term for this is aposematic coloration). These millipedes produce HYDROGEN CYANIDE as a defense mechanism. Yeowch!

You might think that sounds crazy, but in the millipede world, it's just typical. The pink dragon millipede belongs to the largest millipede order, Polydesmida, which consists of about 3,500 species. Millipedes in this order love to produce noxious chemicals – hydrogen cyanide or formic acid – which are released from pores along their body. The liquid either slowly leaks out or sprays out rapidly, depending on the species.

The amount of cyanide that one small millipede secretes is not enough to seriously harm humans (it will burn and blister but you probably won't be in the hospital) but it's enough to kill their predators like birds. If you ever happen to spot one and smell toasted almonds, that's the cyanide.

Ever wondered what the main difference between a centipede and a millipede is? Centipedes have 1 pair of legs per segment of their body, while millipedes have two. The more ya know!

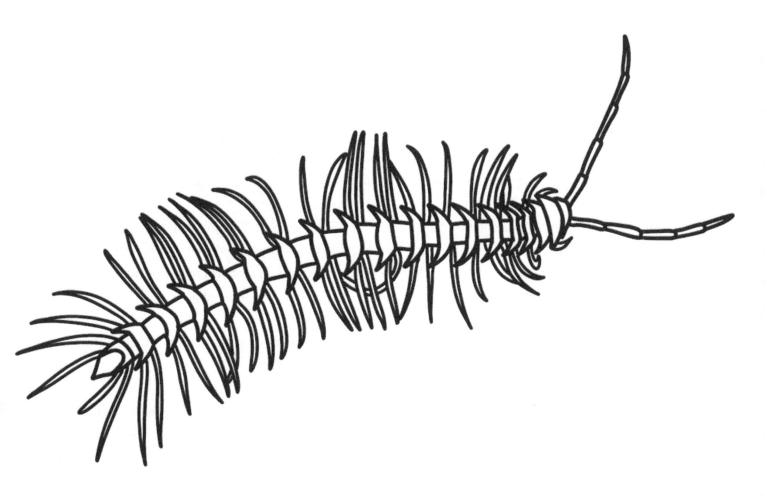

The Blob

Okay, this one is a little different. This organism is very strange. It's not a plant, animal, or fungus. It doesn't have limbs, but it can move. It doesn't have a mouth, but it really likes to eat oats. And it doesn't have a brain, but it can learn, anticipate, and pass on knowledge. Its scientific name is Physarum polycephalum, but it's simply known as "the blob".

The blob is a type of slime mold (NOT a regular mold, which is a fungus – a different grouping). It's classified as an amoebozoa which, truthfully, I don't really know how to describe to you. They're unicellular or multicellular...lumps...I guess? They're able to grow and change shape; the only way I can describe them is like an almost-liquid consistency. The blob is unicellular but can get pretty big.

Scientists have done tons of tests on the blob to understand its level of "intelligence". When multiple parts of the blob are moving (1 cm per hour) in different directions, it will suddenly stop when it runs into itself. So in a way, it "knows" itself. They've also been able to solve a series of mazes, in search of their favorite food (oats). When put into the same maze again, it remembered the shortest path.

Scientists have called the blob a "biological computer". So how does it store information without a brain? Where does it "keep" its intelligence?

Turns out there are vein-like structures all around the body of the blob. They carry nutrients, cellular material, and chemical information all throughout the giant cell. The transportation is visible, as a continuous, pulsing flow. Pretty weird.

And this brings me to why I chose to put a maze on this page. Since the blob can't be defined with a single shape (they are slowly and constantly shifting), I thought it would be wrong to make one up. Instead, the blob can be all of your trial-and-errors and solutions to solving this maze. I mean, that is what the blob had to do to find its oats.

Terror Birds

For nearly 58 million years, the terror birds served as a harsh reminder that not all of the dinosaurs went extinct after the asteroid impact. Around the same time as the appearance of Titanoboa, the terror birds started building their empire in South America. There were species of different speeds and sizes; the largest got to 10 ft (3 m) tall and 1,000 lbs (453 kg), and the fastest could reach speeds of 60 mph (96 kph).

While there was a wide variety of builds, all terror birds had one thing in common: a pickaxe-like hook at the end of their beak. They likely used this hook to strike at their prey repeatedly before devouring it, and it might've helped to tear flesh from the bodies of their victims.

Using fossil evidence, scientists determined that terror birds likely had low-frequency hearing, which they used to track their prey. They also likely communicated with a low, roaring sound (in my opinion, this makes them 10x more terrifying).

South America was an island continent for the majority of the terror birds' reign – up until the very end when it collided with North America. This led to the Great American Interchange, where the terror birds were suddenly met with mammalian apex predators that had been evolving separately for millions of years. They held out for a bit, but this exchange ultimately led to their extinction.

Sea Spiders

You would think someone drew this as an idea of what spiders would look like on another planet. But no. This is lives here. In our oceans.

They're called pycnogonids, but lots of people know them as sea spiders. I mean, they walk the walk (look like spiders) and talk the talk (are arthropods) but they are not true spiders. Otherwise, these would definitely beat the goliath bird eater for the "biggest spider" crown. Deep sea pycnogonids found near the north and south poles can get to 2.5 ft (almost 1 m) long.

There are over 1,300 species and they've been found as deep as 21,000 ft (7,000 m) below the surface. They come in a range of sizes, from a grain of sand to the size of a house cat.

The body of a sea spider is separated into two parts: the cephalon (head) and the trunk (body). And as you've probably noticed, they're literally all leg. They even use their legs to breathe; oxygen seeps through them and into their bodies. They've got so much leg they need to keep their organs in them too. Not all the way out to the tips, but where you can see those first segments end off the trunk. And clearly, this body plan works for them, as they have remained this way for hundreds of millions of years.

Like many other arthropods, pycnogonids have a proboscis to feed (think of a proboscis as a sharp straw). They primarily eat soft-bodied invertebrates like sponges, anemones, and nudibranchs by sucking out their insides. Sometimes their prey is alive once they're done, so in this case, the pycnogonid would be considered a parasite instead of a predator. Larval pycnogonids often live within the tissues of cnidarians (anemones, jellyfish, etc.) as parasites. That makes my skin itch.

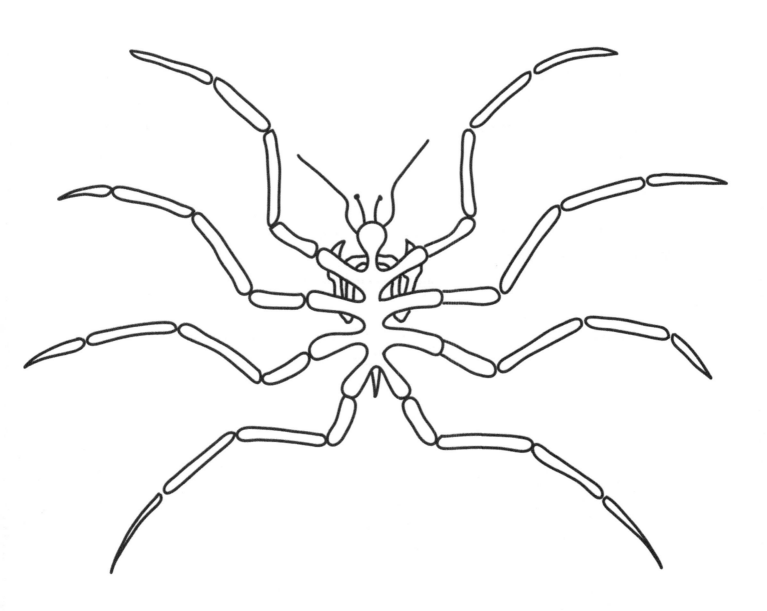

Sea Scorpions

Luckily, these animals are extinct. Eurypterids, commonly known as sea scorpions, were some of the largest arthropods to ever exist (that we know of). They first appeared in the Silurian period over 400 million years ago and dominated the oceans as top predators for 200 million years. Many of them had massive claws that they would use to rip their prey (likely ancient fish and the ancestors to squid) to shreds.

The one pictured here is Jaekelopterus, the largest of all the eurypterids we know of. They could get to over 8.5 ft (2.6 m) long. Fossil evidence suggests they inhabited lakes and rivers, rather than oceans like their relatives.

The eurypterid empire came to an end during the Permian extinction (also known as The Great Dying) 252 million years ago.

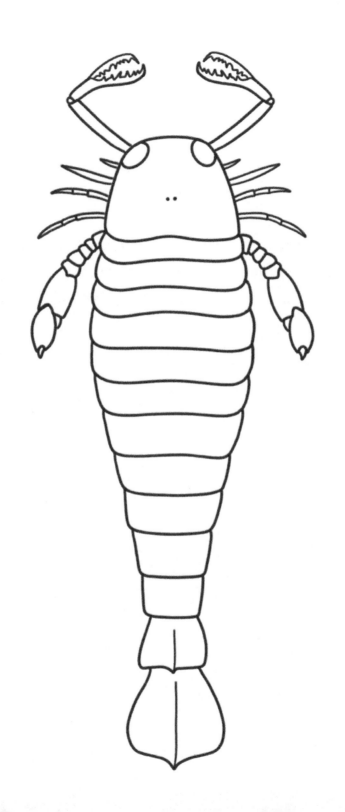

Happy-faced Spider

Whether this one is spooky-cute or just spooky is up to you. I think it's spooky-cute. These spiders are pretty shy, nonvenomous, and kinda just hang out.

An up-close encounter with a happy-faced spider reveals exactly where their name comes from. They each have a unique happy face-like pattern on the top of their abdomen. They come in different colors and some look more like happy faces than others. It's thought that their patterns might prevent them from being eaten by birds. They're really small, too! About 5 mm long. That's the width of a pencil eraser.

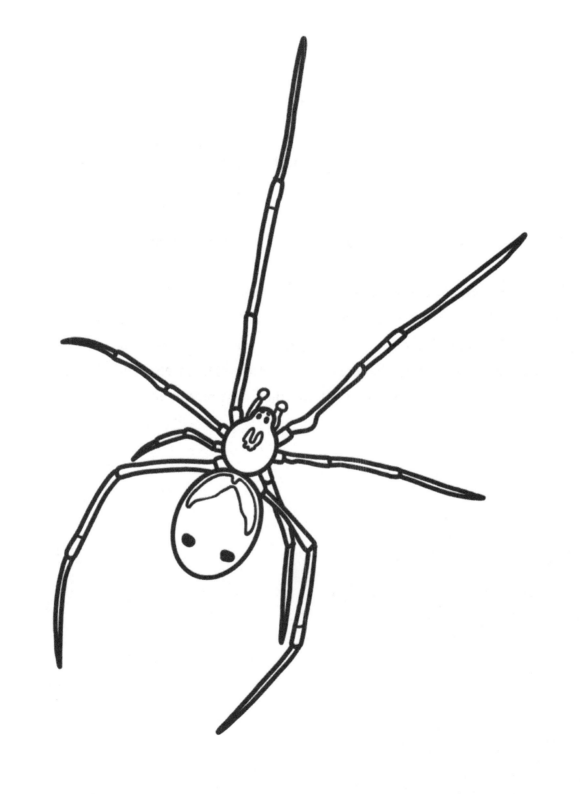

Gulper Eel

I like to call this one the "nightmare balloon from hell". Others call it the gulper eel, the pelican eel, or the umbrella-mouth gulper.

They're found at depths of 1,600 - 9,800 ft (500 - 3,000 m). They get to about 3 ft (1 m) long.

They have extremely small eyes – their eyesight is no good. They also have a whip-like tail, which means they can't swim well either. They kinda just drift around. But as you can see, they have a gargantuan, godforsaken mouth. They can open it wide enough to eat schools of fish or prey animals that are larger than they are. They can also puff themselves up to look like massive balloons to ward off predators (hence my nickname for them).

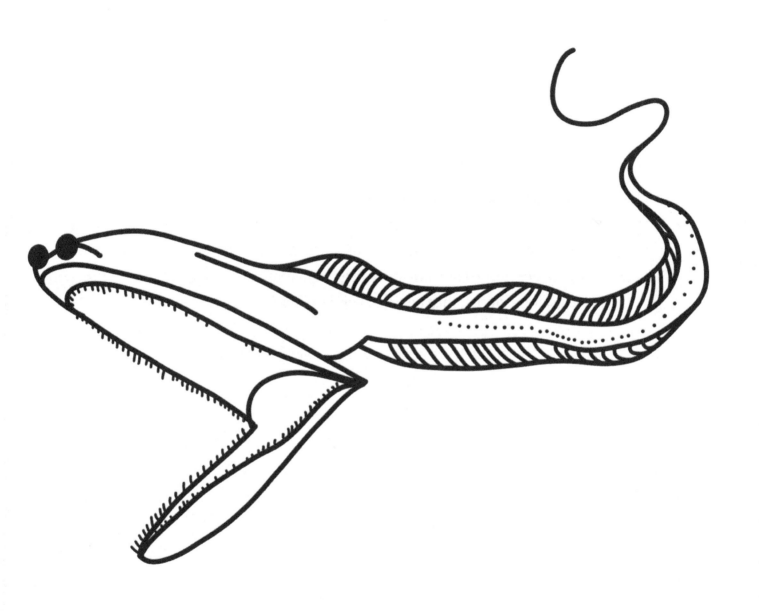

Deep Sea Telescope Fish

The deep sea telescope fish is my son. I have a collage of photos of him framed in my home.
However, I would be lying to myself if I said he didn't belong in this book.
I'm not blind. I know he's spooky.

The telescope fish is up to 8 inches (20 cm) long and found in deeper waters, 1,640 - 6,600 ft (500 - 2,000 m) below the surface. They're named for their tubular eyes that have evolved to suck in all of the light they can get in the depths, as they look upwards towards the surface in such of silhouettes of their next meal. They have extremely flexible jaws to swallow prey larger than themselves, which is then folded in half to fit inside of their stomach.

Unfortunately, not much more is known about them. :(

Viperfish

The viperfish would definitely be in a vacation brochure for Rock Bottom. They have massive fangs that are too large to fit inside their mouths and a bioluminescent light suspended from their heads to lure in prey. That's textbook definition deep sea. It would be weird if I didn't include them.

They're found at depths of 655 - 5,000 ft (200 - 1,500 m) and can get to about a foot long (30 cm). They've been observed hanging motionless with their bioluminescent lure flickering. Spooky. Along with their lure, they have bioluminescent spots on their bellies that make them invisible when viewed from below (this comes in handy with predators that look upwards in search of silhouettes of potential prey, like my son).

Their big teeth aren't for nothing! They can use them to impale prey after swimming at them super fast. :) They can also imprison small fish and shrimp with them after unhinging their jaw and engulfing them.

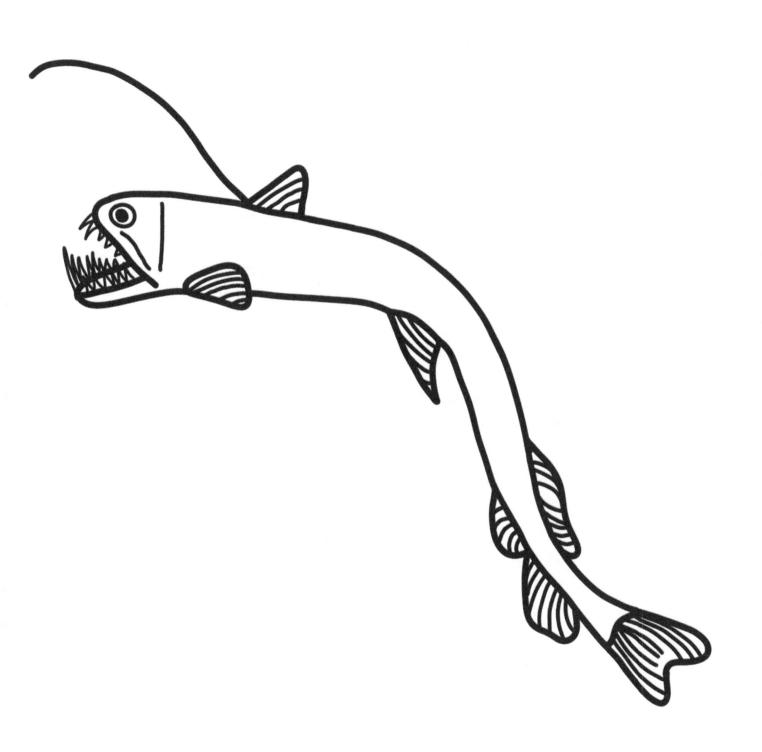

Pipa Toad Back Birth

This is a pipa toad, also known as a pipa pipa. As you can see, they are very flat. As you can also see, there are many holes in the pipa toad's back.
Those are eggs. That's how the pipa toad gives birth.

Pipa toad birth videos are highly regarded as one of the fastest ways to trigger trypophobia (hole-phobia). If you think you can stomach it, go check it out on YouTube.

Or stay here and let me explain it to you!

When it's time to reproduce, male pipa toads will deposit fertilized eggs onto the female's back. The eggs slowly sink into her skin and stay there for 3-4 months as they grow. Once they're ready to enter the world, they eject themselves out of their mother's back – all ready to go as a mini version of their mama and papa. Hundreds of babies can erupt from a female's back at once. And once the process is done, the mother sheds the skin that was used to birth that round of babies.

So the process can begin again.

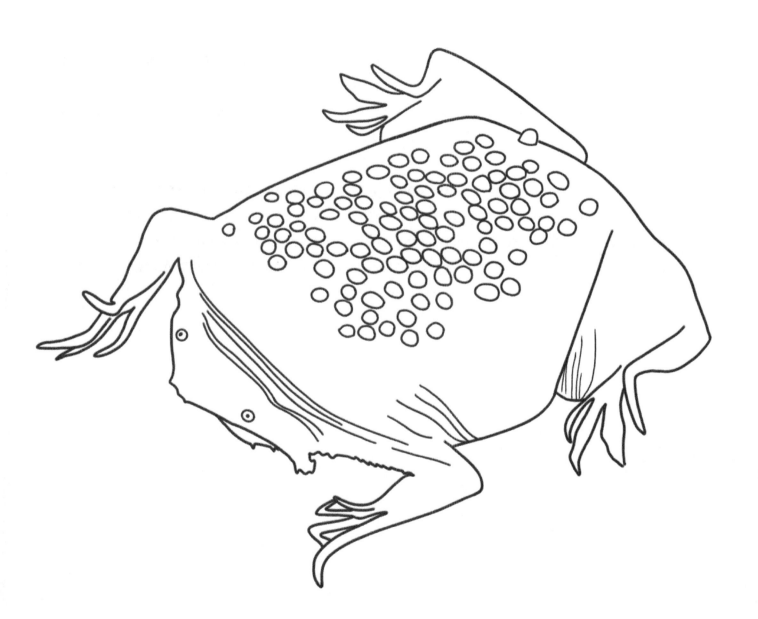

Spider-Tailed Horned Viper

This is another classic case of a bamboozler. Found in western Iran, the spider-tailed horned viper is a species that was only first described in 2006. They're named after their caudal lure: a form of mimicry where a tail evolves to resemble the prey of another organism. In this case, the viper's tail has evolved to look like a spider – the prey of a bird.

Many species use caudal lures to attract prey (sharks, lizards, and other snakes) but I've never seen anything like this one. The first time I watched a video of this viper hunting, I was just as confused as the bird. The viper's tail not only looks like a spider but moves like one too. It almost seems too perfect. It almost seems like evolution "knew" somehow. But it didn't. And that makes it even crazier to think about.

Hallucigenia

One of the coolest extinct animals and one of the coolest named animals, all wrapped up into a cool little dude. Hallucigenia: named for their "bizarre, dream-like quality".

Hallucigenia lived during the Cambrian Period, a time marked by an "explosion" of life that appeared 541 million years ago. This was when most of the major animal groups appeared for the first time in the fossil record.

The Cambrian Period was like nature's trial run. The animals were bizarre, and Hallucigenia was not an outlier. They were 2 cm long and extremely thin, with 7 sets of legs and a matching 7 pairs of spikes. They had 3 sets of appendages near their heads, which they might've used to sense the environment or filter plankton.

It took scientists 14 years to figure out which side of Hallucigenia was up. Initially, they were thought to walk on "stilts", with tentacle-like appendages coming out of their back with mouths at their ends to capture food in the water. But then, new fossil evidence flipped it upside-down! (How you're looking at it right now.) Turns out, the "stilts" were spikes, the "tentacles" were legs, and the "mouths" at their ends were claws!

It took even longer to figure out which side was the head, and this was just determined recently. New fossil evidence found that Hallucigenia had simple eye spots, a mouth with a ring of teeth inside of it, and a row of teeth going down its throat!

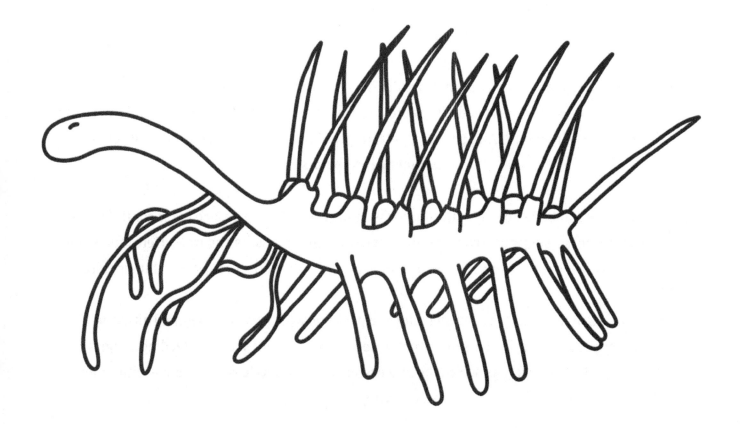

Green-Banded Broodsac

Imagine waking up to find that your eyes have been infected with parasites, and you have no control over what direction you're moving in. That's what happened to this snail. It's been infected with the green-banded broodsac, a parasitic flatworm.

Here's how it happens. An amber snail accidentally eats the flatworm's eggs, which then hatch once they're inside of the snail's body. They grow into structures called "sporocysts" and travel into the eyestalks of the snail. They sit there, pulsating different colors of green, red, and yellow, resembling caterpillars.

Usually, these snails like to stay in the darkness. However, the parasite also takes control of the snail's brain, possessing it to follow the sunlight to the tops of grasses and trees. Here, out in the open with moving caterpillars for eyes, the snail becomes the perfect target for a hungry bird.

The bird SNIPS at the snail's eyes, swallowing the parasite, which is what it wanted all along. This is where the parasite will reach its final form. It settles into the bird's rectum, feeding on passing fecal matter. It then lays eggs, which the bird excretes, and the cycle starts again – with a new, unknowing snail.

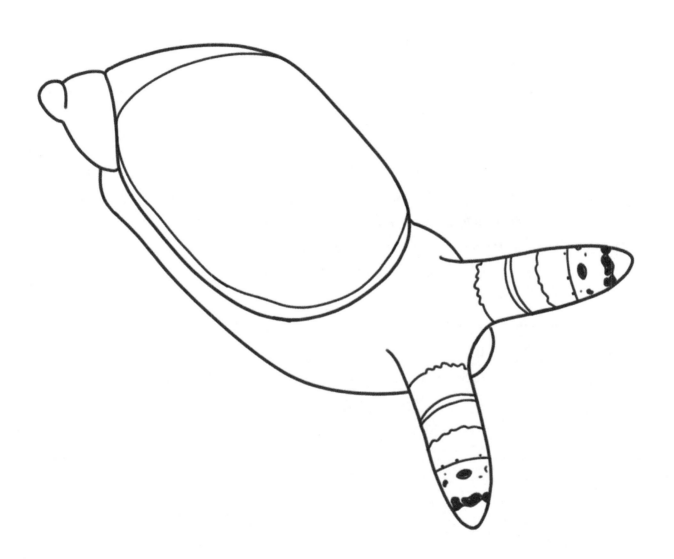

Whip Spider

When I first saw this animal, I thought it was the most demented creature I had ever laid my eyes on. Then I realized, oh! It's because it was in Harry Potter! (During the Goblet of Fire, when Mad-Eye Moody was putting all of those curses on that creature. That was a CGI representation of a whip spider.)

Whip spiders are not true spiders – they're most closely related to whip scorpions (yes, those exist too). There are 150 species found throughout tropical regions, including the southwestern United States and Florida. They have 4 pairs of legs, with the first pair elongated to look like whips. They use pedipalps (which are like jaw claws) to capture prey such as grasshoppers, roaches, crickets, and even small birds and lizards.

But they're really not as bad as they look. According to an ecologist, they are "shy, timid, and very delicate." They're known to care for their offspring for many weeks, and potentially months, and are very territorial with each other. They're safe to observe from a distance (I get it if you don't want to though.)

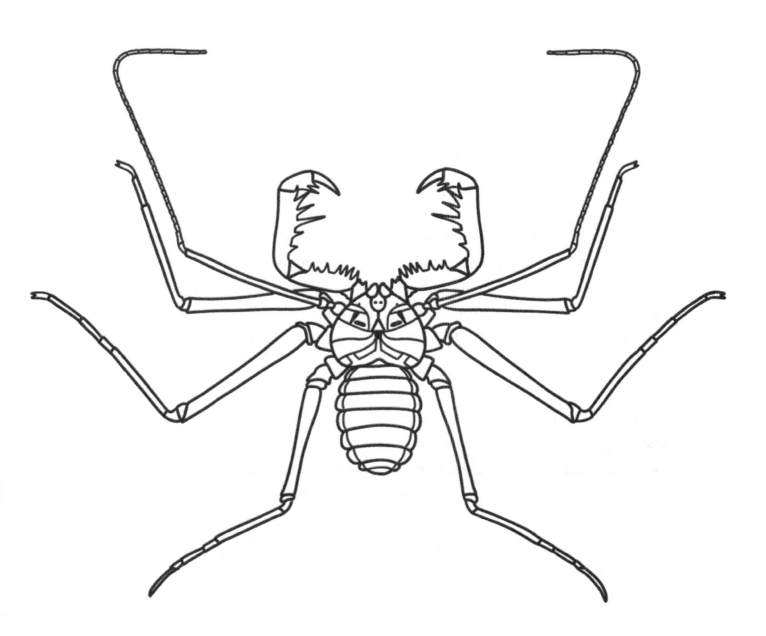

Hagfish

The Smithsonian calls the hagfish the most disgusting animal on Earth. Initially, this may seem a bit harsh. But once you read through their facts, I hope you understand why they're objectively repulsive.

The hagfish is a jawless and boneless fish. There's over 70 species and they've been known to live as deep as 5,500 ft (1,700 m) below the surface.

A closer look at their face reveals sensory tentacles around their mouth, used for seeking out the flesh of their next meal. Their eyespots are only able to detect light, so they're no good for this purpose. Inside of their mouth is an unusual arrangement of teeth: two rows, unbraced by jaws, arranged vertically on their face. It's visually unpleasant. They use these teeth to pummel into carcasses, eating their way through tunnels of flesh as they consume them from the inside out.

Their worm-like body is capable of an arguably disrespectful defense mechanism: producing lots and lots of slime. The slime expands rapidly in water, and can fill a bucket instantaneously. A potential predator, in search of a quick meal out of the hagfish, instead receives a mouthful of slimy liquid. Imagine that. Even the absence of food, they're able to absorb nutrients through their skin and can live for months this way.

They've been around for over 300 million years and have experienced very little change to their body plan. This somewhat implies that they've reached an ideal form, but that's not something that I want to consider.

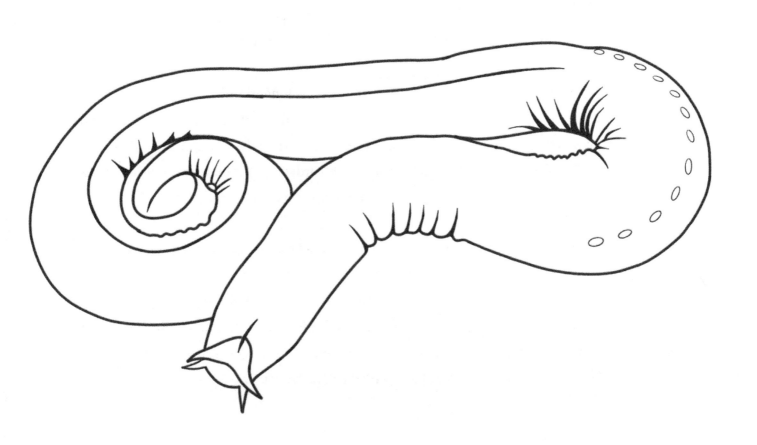

Ribbon Worm

It's almost impossible to put the ribbon worm into a single descriptive category. There's over 1,000 species that come in a variety of sizes: ranging from 1 cm long to over 98 ft (that we've recorded). They have the potential to hold some of the longest animals on Earth.

The "potential" part might be confusing. If they could be the longest animals, why don't we know for sure? Well, one thing ribbon worms all have in common is their stretchiness: they can expand to swallow prey wider than their bodies, and can contract to 1/10 of their length when threatened. Their size is fluid. So, the longest ribbon worm ever recorded at 98 ft has the potential to have actually reached over 160 ft long. We just don't know for sure.

I want to say more on their defense mechanism. I mentioned they contract to a fraction of their size in the presence of a threat. They also push out their proboscis, which means something different for worms than for spiders/arthropods. Rather than a straw-like appendage, the proboscis of a worm is more like branching spaghetti that can have different properties, depending on the ribbon worm in question. Some have sticky ones, others are spikey, and some even eject toxic chemicals. So don't mess with a ribbon worm.

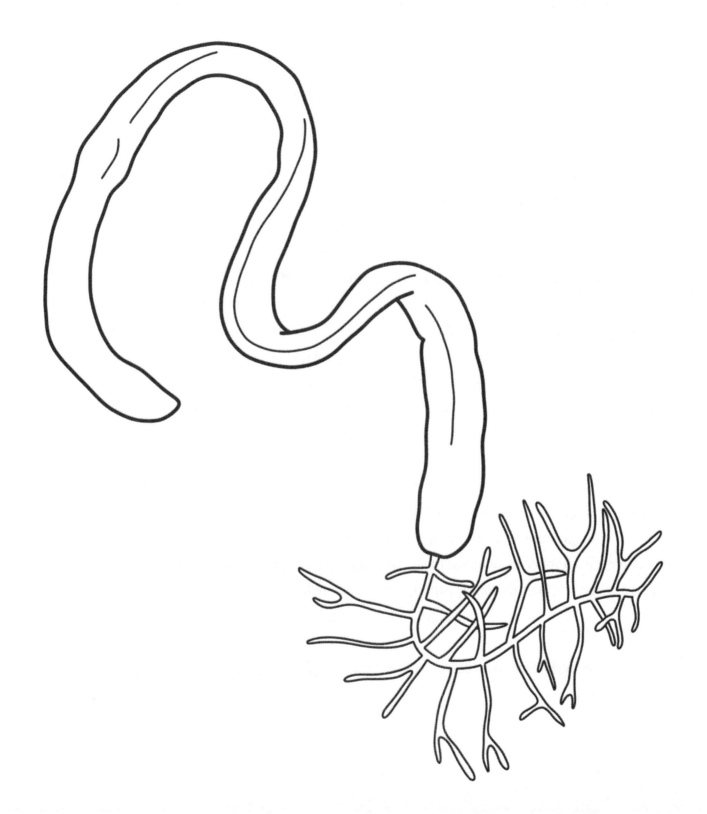

Black Mamba

This snake is a creature from my own nightmares, as it's the only animal that has ever truly scared me. They're found in Sub-Saharan Africa, including Namibia, where I went to intern at the Cheetah Conservation Fund back in 2019. Before arriving, I had never heard of the black mamba (at the time, I didn't pay much attention to reptiles) and I was shocked by a harsh reality when I got there.

Early on, I found out that the black mamba is one of the most venomous snakes in the world, and was frequently spotted in the part of the African bush where I was living. I found out that their venom can kill you in 30 minutes and without antivenom, fatality from a bite is 100% certain. The site where I was living happened to be an hour away from the nearest hospital. I asked my supervisor, "So, what happens if I get bit by a black mamba? What will we do?" in hopes of a comforting answer. She replied, "I'd give you your phone so you could text your friends and family goodbye."

Let me give you a clear picture of what they look like. They're not named for the color of their scales, as many people assume. They get their name for the color of the inside of their mouth, which they display as a warning before one might meet their fate. They can get up to 14 ft (4.2 m) long. They also happen to be one of the fastest snakes in the world, reaching speeds of 12 mph (19 kph) – a speed you most definitely cannot reach. They're also able to move quickly with ⅓ of their body lifted off the ground...heinous.

Luckily, I never encountered one, but the thought of them kept me up at night.

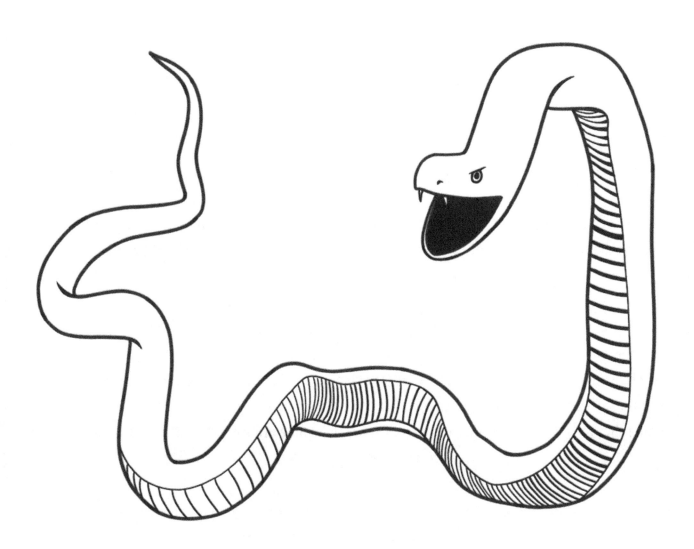

Rat King

This was the last picture to get finished for this coloring book. That's probably not a surprise to you. It's unpleasant to look at, isn't it?

This is a rat king. And it's exactly what it looks like. Multiple rats tangled together at the tails. It happens on VERY rare occasions – only about 60 rat kings have been seen in the last 500 years – and has ranged from 3 - 30 rats tied up together in this mess. The only way out of it is death.

So why is it called a rat "king"? It's supposed to represent the evilness and filth of royalty, and how the king takes advantage of the lower class or something? I don't know. I'm not a historian. Or an English major.

I can tell you how this happens, though! No one has ever seen one develop (and I would hope that someone would step in to help if one started to) but it seems to be clear how and why it occurs. It must happen in confined, small spaces during cold months where rats are bundling together for warmth. Since their tails have a grasping reflex, it's easy for them to get tangled up together. The longer they stay there, the more substances get involved (like urine and blood) which make the knot too sticky to untangle.

For a long time, it was thought that rat kings were manmade until a living rat king was spotted for the first time, just last year, in Europe!

Hope You Enjoyed this Coloring Book!